BEI GRIN MACHT SICH IHR WISSEN BEZAHLT

AF151435

- Wir veröffentlichen Ihre Hausarbeit, Bachelor- und Masterarbeit

- Ihr eigenes eBook und Buch - weltweit in allen wichtigen Shops

- Verdienen Sie an jedem Verkauf

Jetzt bei www.GRIN.com hochladen und kostenlos publizieren

Yumna Mubashir

ELISA (Enzyme-Linked Immunsorbent Assay) Practical Report

GRIN Verlag

Bibliografische Information der Deutschen Nationalbibliothek:

Die Deutsche Bibliothek verzeichnet diese Publikation in der Deutschen National-
bibliografie; detaillierte bibliografische Daten sind im Internet über http://dnb.d-
nb.de/ abrufbar.

Impressum:

Copyright © 2015 GRIN Verlag GmbH
Druck und Bindung: Books on Demand GmbH, Norderstedt Germany
ISBN: 978-3-656-95722-5

Dieses Buch bei GRIN:

http://www.grin.com/de/e-book/299024/elisa-enzyme-linked-immunsorbent-assay-
practical-report

GRIN - Your knowledge has value

Der GRIN Verlag publiziert seit 1998 wissenschaftliche Arbeiten von Studenten, Hochschullehrern und anderen Akademikern als eBook und gedrucktes Buch. Die Verlagswebsite www.grin.com ist die ideale Plattform zur Veröffentlichung von Hausarbeiten, Abschlussarbeiten, wissenschaftlichen Aufsätzen, Dissertationen und Fachbüchern.

Besuchen Sie uns im Internet:

http://www.grin.com/

http://www.facebook.com/grincom

http://www.twitter.com/grin_com

ELISA PRACTICAL REPORT

Introduction

The Enzyme-Linked Immunosorbent assay (ELISA) is a test that is conducted to identify the substance under study with the help of antibodies and changing color elements. It is the "wet-lab" oriented analytic biochemistry assay that consume enzyme immunoassay to detect the existence of a substance (antigen) in a liquid sample. It is of great help in medicine and plant pathology as a diagnostic tool and quality control check.

ELISA test involves one antibody according to the need of the test, where the antigen is immobilized on the solid plate. The method is done specifically and nonspecifically. The report involves the description and construction of two different ELISAs.

Theory behind Fibrinogen

It is possible to differentiate between the intact and processed Fibrinogen using Sandwich ELISA. In theory, the sandwich ELISA is an efficient method to detect sample antigen. The method involves the two layers of antibodies i.e. detection and capture antibodies. An antigen under study must have at least two antigenic epitope that are capable to bind an antibody (monoclonal and polyclonal) and hence easily detectable in the system (ELISA encyclopedia).

The biochemistry of fibrinogen can be manipulated when designing an appropriate ELISA sandwich assay. On the other hand, intact fibrinogen is only present in plasma. Fibrinogen is a protein produced by the liver that helps to stop bleeding by forming blood clots. During normal clotting, Fibrinogen is broken by thrombin that is also an enzyme, into short sections of fibrin. It can be measured in serum or plasma (Brunner, E.; Smith, G.; Marmot, M.; Canner, R.; Beksinska, M.; O'Brien, J.).

Sandwich ELISA is used to distinguish both Fibrinogen because it utilizes two antibodies to detect. The first antibody is the capture, and other is detection. These two antibodies detect the distal portion of intact fibrinogen that in turns useful to distinguish plasma and serum.

Theory of Gliadin

In this method, the single antibody is used and hence the method is also called Elisa. The theory behind the ELISA is that the single antibody is directed against the antigen that has to be immobilized on the surface. Single antibody makes it simple to detect depending upon the protein.

Gliadin is the complex group of protein that coupled with the gluten protein to determine the properties of wheat flour. It also helps the bread to rise in the process of baking. It is important to study the wheat flour quality.

On the other hand, this method is used to detect the childhood coeliac disease and adult coeliac disease and are preferable for screening of coeliac patients.

AIM

There are basically two major sections of the experiment, therefore, for each of these sections the outcomes and aims are different. The experiment holds two different section, in each section ELISA is conducted on human fibrinogen and gliadin proteins. These two proteins from different sources are taken. Hence, these two methods (repeated for one another) holds the objectives respectively.

The Chief aim of the experiment is to establish the ELISA Standard Curve from the provided sample that contains a specific concentration of human fibrinogen and gliadin proteins. Fibrinogen is extracted from the human serum whereas, gliadin is taken from bread. Moreover, the study is also conducted to determine the concentration of above-mentioned protein in unfamiliar tester by using standard curves. Lastly, the experiment is conducted to compare the single and double Ab ELISA methods.

Methods

The method of both experiments is same. However, protein Ag is changed in plate 1 and plate 2. The experiment started with the adding 100 µl buffer in the A1, A2, B1 and B2 funnels. These funnels are blank. The Same amount of buffer is also added in the A4-A12, B4-B12, C1-C12 and D1-D12. On the other hand, A3 and B3 are the bells having the 100 µl Ag protein (either fibrinogen or Gliadin). An undiluted amount of antigen is added to it. Remaining wells have same series in which 100 µl of stock of protein is added in A4/B4 with a gentle mix of 100 µl coating buffer. Therefore, these wells contain 200 µl liquid and dilution ratio is 1:2.

2

The next step is to add 100 μl liquid in A4 to 100 μl buffer solution in A5 and stir. The dilution ratio is doubled as previous 1:2 which is now 1:4. Now the dilution of the stock top standard is 1:4. The remaining two well will remain same at the point. The next step is to mix the A5 well, which contained 100 μl into the next well that is A6 containing 100 μl buffer coating. Now the dilution of the stock is 1:8. Well, 7 remain same at the point.

The same method is repeated for all the rows in A section and continued till Row C. at the time when, when experiment of adding reach to the well C12, mix the 100 μl that is freshly transferred with the 100 μl having coating buffer and later dump the solution of 100 μl into waste. It is done to have the same amount of liquid reagent in row A and row C. after discarding the last 100 μl, row A, and row C now have 100 μl liquid reagent in every well.

Above mentioned steps are repeated for row B and row D. the duplication of the procedure gives out the standard curve. Here one set in row A plus row C and one set in row B and row D. in other words, B4 is the duplication of A4 and D1 is the duplication of C1. It must not be confused with the rows because A and C have concentration so do B and D.

Lastly, three different unknown 100 μl human serum extracts are added per well to plate 1 labeled as FX, FY, and FZ. The plate must be covered with the adhesive film and incubate at 4°C until for part 2.

The same experiment is repeated for the unknown extract of bread (gliadin) in plate 2 labeled as GX, GY, and GZ).

Results

The data was collected through carefully through the given protocols and results thus found are as follows.

Fibrinogen Result:

	1	2	3	4	5	6	7	8	9	10	11	12
A	0.337	0.256	2.591	2.895	2.944	2.878	2.499	1.572	0.866	0.500	0.358	0.269
B	0.395	0.289	2.764	2.960	2.919	2.800	2.246	1.022	0.708	0.388	0.304	0.232

C	0.292	0.315	0.292	0.149	0.114	0.158	0.159	0.128	0.218	0.133	0.201	0.177
D	0.233	0.226	0.187	0.176	0.200	0.223	0.195	0.176	0.167	0.149	0.139	0.142
E	2.309 (X)	0.498 (Y)	0.225 (Z)									
F	2.814 (X)	0.693 (Y)	0.260 (Z)									

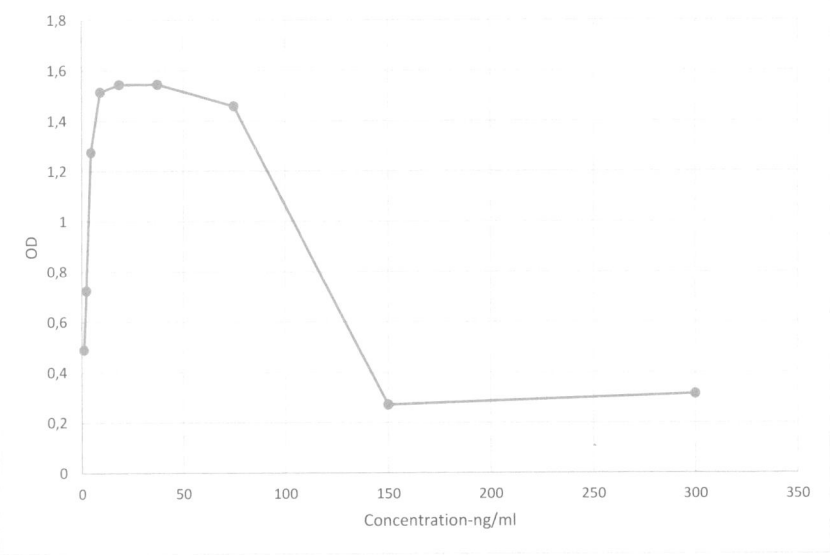

Fibrinogen Result

The equation $(Y=0.016X + 0.489)$ allows you to translate, and co-relate the OD with the relative value of concentration. The value of R^2 indicates how much the data is correct which in our case is nearly equal to 1(which is perfect)

Any calculation can be easily done with the data and the charts. If the OD values which are in ng/ml are diluted too much then multiply it by the suitable factor to get the original values. The x-intercept and the y-intercept can also be calculated by putting x=0 and y=0. To make the readings easier Excel allows you to make the non-linear graph a section of the standard graph by adding a trend line. The equation will become in the form of $Y=AX^2 +BY+C$. it can be dragged and pasted to other cells.

Gliadin Result:

	1	2	3	4	5	6	7	8	9	10	11	12
A	0.407	0.348	1.217	1.217	1.292	1.375	1.324	1.154	0.733	0.921	0.617	0.392
B	0.347	0.382	1.416	1.383	1.095	1.265	1.076	1.004	0.784	0.504	0.379	0.432
C	0.428	0.299	0.241	0.197	0.428	0.247	0.372	0.302	0.306	0.301	0.250	0.552
D	0.607	0.539	0.533	0.520	0.438	0.491	0.506	0.180	0.202	0.409	0.289	0.261
E	0.589 (X)	0.257 (Y)	0.191 (Z)									
F	0.699 (X)	0.282 (Y)	0.250 (Z)									

Gliadin Result:

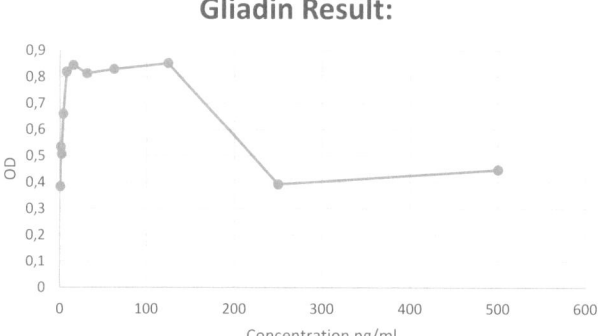

The equation (Y= -0.03X + 0.38) allows you to translate, and co-relate the OD with the relative value of concentration. The value of R^2 indicates how much the data is correct which in our case is nearly equal to 1(which is perfect)

Any calculation can be easily done with the data and the charts. If the OD values which are in ng/ml are diluted too much then multiply it by the suitable factor to get the original values. The x-intercept and the y-intercept can also be calculated by putting x=0 and y=0. To make the readings easier Excel allows you to make the non-linear graph a section of the standard graph by adding a trend line. The equation will become in the form of $Y=AX^2+BY+C$. it can be dragged and pasted to other cells.

Discussion

The main of this project was to make Elisa Standard Curve from the provided data that was related to the specific concentration of human fibrinogen and gliadin proteins. Also, another theme was to perform tested (Excel) data from the Elisa Standard Curve. All the requirements were met during the experiment and data analysis, and there was no value that was unexpected or certain. Accurate experimental data led to perfect Elisa Standard Curve due to immense hard work and collaboration of the technicians. If something occurs, the reason for that is the inaccurate data of formulae during the analysis.

By looking at the analysis of my experiment and the standard curve obtained, I come to the conclusion that both the Single Body (monoclonal) and the Double body (polyclonal) are essential and both have their own consequences and leads. The Polyclonal antibodies are very inexpensive and user-friendly, even beginners can easily opt to that whereas the monoclonal or single antibodies are rather a little expensive when compared to the double antibodies. When performing experiment if the desired time scale is precise and less than it is better to opt to Polyclonal antibodies because it is less time-consuming. But on the other hand if your time frame is a little big then monoclonal antibodies require a little more time for hybridism. Both antibodies are good to use with your own requirements. If your experiment requires specified epitope then it beneficial to use single antibodies. As far as the polyclonal antibodies are concerned it has some up hand advantages that due to the fact it can enhance the target signal for proteins even at low levels due to its high affinity. It also gives better results in IP/ChIP. Denatured compound or polymorphism doesn't alter the results in double antibodies but makes a great deal of effects in single antibodies. However, it may have some consequences like it double antibodies are much prone to specific batch variability. A large no. of unnecessary antibodies are produced which hinders the affects the final results. It is not handy when it comes to probing specific domains because the multiple domains are recognized by the antiserum. . So overall double or polyclonal antibodies are preferred more (Abcam).

Another accurate method for measuring protein levels/concentrations is the Bradford Assay since accurate protein levels are the key requirements in any experiment. This works simply because it is based upon collaboration between Coomassie brilliant blue that is used for staining gels and aromatic and arginine residue in your protein sample. Maximum adsorption from 470 to 595 nm is required which is gained when the dye impasses the residue. When the standard curve is made, the concentration of protein is calculated on the basis of absorbance. The Bradford Assay is useful because it is fast and interference of reducing agents. But, however, the SDS can interfere with the dye's ability to bind the residue (Cable).

There are numerous applications of the ELISA in the real world when it comes to the identification of antigen or an antibody in the sample. A known application would be the determination of serum in an antibody in a virus test. Most popular are the HIV and the Ebola. ELISA is widely applied in home pregnancy test and while keeping a check on food adulteration in peanuts, almonds, milk, walnuts and eggs. It is also used for screening of drugs (Sino Biological Inc.). A very common treatment and medicine of fibrinogen/gliadin ELISA is the Celiac disease which is the result of the inaccurate response of the immunes to gluten that is the protein found in wheat and barley. Fibrinogen/ gliadin ELISA assays are developed to monitor the disease and suggest a suitable diagnosis. One of the diseases in which ELISA helped the most is the famous HIV. Millions are people are screened out for HIV due to ELISA experiment. The test is simple and easy. The doctor will take the sample of your blood and add it to the petri dish. If your blood contains the antibodies then the two will bind, and you will probably be suffering from HIV and other viruses (Kinman).

Elisa has been the top primary methods for analytic detection for a decade, and several new advancements have been made in few years. The traditional colorimetric must be upgraded by the use of new technologies. New methods employ use particle-based or bead components that are useful for multiplexing. New kits for ELISA are also in the market, and when combined with the breadth of existing module it can be employed on various new expertise (Comley).

In the end I want clinch in the discussion that by considering the advantages and disadvantages of single and double bodies when it comes to analysis, the double body or polyclonal antibodies must be favored while doing experiment in ELISA. Due to which the results are more correct and specific, and errors can be easily omitted.

The equation (Y=0.016X + 0.489) allows you to translate, and co-relate the OD with the relative value of concentration. The value of R^2 indicates how much the data is correct which in our case is almost nearly equal to 1.

Any calculation can be easily done with the data and the charts. If the OD values which are in ng/ml are diluted too much then multiply it by a suitable factor to get the original values. The x-intercept and the y-intercept can also be calculated by putting x=0 and y=0. To make the readings easier Excel allows you to make the non-linear graph a section of the standard graph by adding a trend line. The equation will become in the form of $Y=AX^2 +BY+C$. it can be dragged and pasted to other cells.

Bibliography

Abcam . "A comparison between polyclonal and monoclonal." (n.d.).
 <http://www.abcam.com/protocols/a-comparison-between-polyclonal-and-monoclonal>.

"Brunner, E.; Smith, G.; Marmot, M.; Canner, R.; Beksinska, M.; O'Brien, J." *Childhood social circumstances and psychosocial and behavioural factors as determinants of plasma fibrinogen* 347 (1996): 1008-1013.

Cable, Jennifer. "How to Measure Protein Concentration More Accurately." (2011).
 <http://bitesizebio.com/10178/how-to-measure-protein-concentration-more-accurately/>.

Comley, John. "ELISA Assays: recent innovations take analyte detection to new levels. ."
 (2012). <http://www.ddw-online.com/screening/p191009-elisa-assays:-recent-innovations-take-analyte-detection-to-new-levels-fall-12.html>.

ELISA encyclopedia . *Sandwich ELISA, Highly Sensitive.* 2012. <http://www.elisa-antibody.com/ELISA-Introduction/ELISA-types/sandwich-elisa>.

Kinman, Tricia. "ELISA." (2012). <http://www.healthline.com/health/elisa#Procedure2>.

Sino Biological Inc. "ELISA Applications / Applications of ELISA." (n.d.). <http://www.elisa-antibody.com/ELISA-applications>.